YOUR KNOWLEDGE HAS VALUE

Mohamed Etman et al.

Effect of Zinc Oxide Nanoparticles on Listeria Monocytogenes in Ready-to-Eat Meat

GRIN Publishing

Bibliographic information published by the German National Library:

The German National Library lists this publication in the National Bibliography;
detailed bibliographic data are available on the Internet at http://dnb.dnb.de .

Imprint:

Copyright © 2015 GRIN Verlag GmbH
Print and binding: Books on Demand GmbH, Norderstedt Germany
ISBN: 978-3-656-94631-1

This book at GRIN:

http://www.grin.com/en/e-book/298350/effect-of-zinc-oxide-nanoparticles-on-listeria-
monocytogenes-in-ready-to-eat

GRIN - Your knowledge has value

Since its foundation in 1998, GRIN has specialized in publishing academic texts by students, college teachers and other academics as e-book and printed book. The website www.grin.com is an ideal platform for presenting term papers, final papers, scientific essays, dissertations and specialist books.

Visit us on the internet:

http://www.grin.com/

http://www.facebook.com/grincom

http://www.twitter.com/grin_com

EFFECT OF ZINC OXIDE NANOPARTICLES ON LISTERIA MONOCYTOGENES IN READY TO EAT MEAT

(Ebeed,A.Saleh [1]., ELsayed,E.Hafez[2].,Safaa,H.Gorbal[3] and Etman,M.M [4])

----

[1] Department of Food Hygiene, Faculty of Veterinary Medicine, Damanhour
University, Egypt .
[2] Department of Plant Protection and Bio nuclear Diagnosis, City of Scientific
Researches and Technological Applications, Alex, Egypt .
[3] Damanhour Directorate of Veterinary Medicine ,General Veterinary Authority,
Ministry of Agriculture, Egypt .
[4] Meat inspector in Ghayathi slaughterhouse, Western Region Municipality ,
Abu-Dhabi, U.A.E .

Abstract :

The purpose of this study was investigation of the effect of zinc oxide nanoparticles
(Zn O NPs) on Listeria monocytogenes in Ready-to-eat meat products, to introduce a
new, cheap, safe and fast way of food preservation. This work was performed on 240
Random Samples of RTE meat products (Frankfurter, Salami, Basterma & Luncheon)
(60 sample of each product). there was a significant increase in Listeria
monocytogenes CFU/g in the 1st control group of each product which was not
treated by any concentration of Zn O NPs, while in the 2nd group of each product
which was treated with 60 ppm of Zn O NPs, there was some inhibition of CFU/g ,
and in the 3rd group of each product which was treated with 90 ppm of Zn O NPs,
there was a significant inhibition which is matching with the highest concentration of
Zn O NPs used in this study .

Keywords : Listeria monocytogenes, zinc oxide nanoparticles, RTE meat.

Introduction:

Food borne diseases are a substantial public health concern worldwide. Both developed and developing world suffer from severe food borne illness consequences, but to a variable extent **(King et al., 2000)**. the U.S Centers for disease control and Prevention (CDC) estimate food borne outbreaks cause 76 million illnesses and about 5000 deaths annually in the United States **(Mead et al., 1999)**. **Scharff (2010)** provided recent estimates that the cost of food borne illness in the U.S Accounted for 152 billion Dollars each year.

Food borne infections can cause several illnesses in the general population including healthy adults. However, older adults (those who are over 60 years old) tend to have more severe complications to these infections. Also, elderly persons are more susceptible to food borne illness infections and deaths **(Buzby, 2002)**. According to **Smith (1998)**, older adults are more likely to experience severe illness and deaths from gastroenteritis than younger adults.

Listeria monocytogenes causes listeriosis, a rare disease, but it is potentially serious **(Allerberger&Wagner, 2010)**, the average case fatality rate of human listeriosis is from 20% to 30% **(Swaminathan&Gerner-Smidt, 2007)**.

Listeria monocytogenes can be found everywhere in the environment, in soil, water, sewage, silage and fecal materials. Soil is commonly known as a harborage of the bacterium, particularly in silage **(Bourry&Poutrel, 1996)**. Also, it has been reported that a wide variety of healthy animal species shed Listeria in their feces. The stressful factors that animals encounter during transportation to slaughter plants has been reported to increase the shedding of Listeria. In addition, symptomatic and a symptomatic humans do excrete the bacterium in their feces.

Among the various metal oxides studied for their antibacterial activity, zinc oxide nanoparticles have been found to be highly toxic as antibacterial. Moreover, their stability under harsh processing conditions and relatively low toxicity to human combined with the potent antimicrobial properties favors their application as antimicrobials .

According to another study by **(Jin et al., 2009)**, ZnO quantum dots, and nanoparticles of purified powdered ZnO, were effective in reducing the cell populationof Listeria monocytogenes, Salmonella Enteritidis, and E. coli O157:H7. The antimicrobial efficacy was concentration dependent, as higher ZnO concentration resulted in a greater

2

reduction of growth. To be specific, 3.2 mg ZnO/mL treatment caused a 5.3 log reduction of L. monocytogenes and a 6.0 log reduction of E. coli O157:H7 in growth media after 2d incubation. ZnO at 1.12 and 0.28 mg/mL concentrations were investigated against Salmonella for antibacterial effects and as a consequence, cell growth was reduced by 6.1 and 4.1 log CFU/mL, respectively. the results demonstrate the antibacterial activity of ZnO NPs over a spectrum of bacteriaThis was further supported by a study conducted by (Jones et al.,2008), where ZnO NPs were tested for their inhibitory effect on various bacteria including Staphylocoocus aureus, Staph. epidermidis, Staph. pyogenes and Bacillus subtilis. The growth reduction was greater at higher ZnO concentrations and/or smaller particle size, suggesting that the antibacterial mechanism of ZnO nanoparticles against bacteria cells was through accumulation of ZnO NPs inside the bacterial cell membrane.

Moreover, lower concentrations were used in a study by (Liu et al., 2009) where ZnO NP suspensions were analyzed at concentrations of 3, 6, and 12 mM specifically against E. coli O157:H7. Results showed that tryptic soy agar (TSA) plates with 3 and 6 mM ZnO nanoparticles exhibited less bacterial growth over that of a control while at 12 mM, the growth of E. coli O157:H7 was completely inhibited. The researcher also suggested a mode of action of ZnO NPs, which was similar to that by Jones (2008) that ZnO NPs damaged the bacterial cell membrane and caused a lost of intracellular contents.

Materials & methods :

1. Collection of samples (ISO-17604:2003):

A total of 240 random samples of ready-to-eat meat products (Frankfurter, Salami, Basterma & Luncheon) (60 samples of each product) were collected from different supermarkets and shops In Al-Behira and Alexandria governorates. Each sample weighted 250 gm; the collected samples were transferred in an insulated ice box under complete hygienic conditions rapidly and directly as possible to the Accredited Food and Feed Safety Laboratory, Faculty of Veterinary Medicine, Damanhour University.

2. Apparatus and instruments: (ISO 7218:2007)

Microbiological laboratory equipment's according to (ISO 7218:2007).

3. Chemicals

- Distilled water was used to prepare the solutions and to wash the glassware in each experimental run.
- Zinc Acetate dihydrate Zn (CH3COO).2H2O (purity 99%) (RANKEM, India).
- Sodium Hydroxide NaOH (purity 99%) (SIGMA-ALDRICH, Germany).
- Poly Vinyl Pyrolidone (PVP) (C6H9NO) n (purity 95%) (EASTERN FINE CHEMICALS, Italy).
- Absolute Ethanol C2H5OH (purity 99.8%) (SIGMA-ALDRICH, Germany).

4. Listeria Reference Strain

Selectrol discs are intended for use in microbiological laboratories for the control of test methods. Being 1st generation traceable to vials of recognized national type culture strains, Selectrol discs are acceptable in accredited laboratories for the production of working stock cultures.(Lot no. 48/52   Buckingham MK 182 LR UK.)

Selectrol discs are freeze-dried microbial preparations manufactured from NCTC (National Collection Of Type Cultures) And NCPF (National Collection Of Pathogenic Fungi) Cultures

5. Selective primary enrichment medium (Half Fraser broth)(Oxoid)

Table (1): Composition of Half Fraser broth

Material	Concentration
Meat peptone (peptic digest of animal tissue)	5,0g
Tryptone (peptic digest of casein)	5,0g
Beef extract	5,0g
Yeast extract	5,0g
Sodium chloride	20,0g
Disodium hydrogen phosphate dehydrate	12,0g
Potassium dihydrogen phosphate	1,35g
Aesculin	1,0g
Water	1000ml

6. Agar Listeria according to Ottaviani Agosti (ALOA)

A special isolation medium which is not only selective for listeria but which allows the direct differentiation of L. monocytogenes in the presence of other Listeria and other background flora was evaluated. Agar Listeria according to Ottaviani and Agosti (ALOA) is both a selective and differential medium for the isolation of Listeria spp. and presumptive identification of

L. monocytogenes **(Ottaviani ., et al 1997a)** , **(Ottaviani ., et al 1997b)** The selectivity is obtained by adding a series of antimicrobial compounds comparable to the PALCAM medium. The indicative character is established by the introduction of chromogenic substrates in the selective medium. Chromogenic substrates have proven to be a powerful tool, utilizing specific enzymatic activities of certain micro-organisms **(Manafi 1996).** Improved accuracy and faster detection of target organisms can be obtained. They have been in use for some years and are at present most commonly used for the detection and enumeration of coliforms, Escherichia coli and Enterobacteria **(Manafi ., et al 1991).**Some applications have also been described for Salmonella, Clostridium perfringens, Staphylococcus aureus and Lactobacillus acidophilus**(Kneifel& Pacher,1993),(Kühn .,et al 1994).**

They have recently been introduced for the isolation and/or detection of L. monocytogenes. In the ALOA medium, the chromogenic compound X-glucoside is added as substrate for the detection of β-glucosidase, which is common for all Listeria species. The differentiation of L. monocytogenes from other Listeria spp. is based on the production of a phosphatidyl inositol specific phospholipase C in L. monocytogenes strains which can hydrolyse the specific purified substrate added to the medium, resulting in an opaque clear cut halo surrounding the L. monocytogenes colonies.

Methods

1. Synthesis of Nano-zinc Oxide via Sol-gel Technique:

Zinc oxide Nano particles powder with different morphological structures was synthesized using sol-gel technique. The variation of the preparation conditions was optimized in order to attain ZnO Nano powder with good morphological structures according to the following procedures:-

A.   14 mm aqueous solution of Zinc acetate dihydrate Zn $(CH_3COO).2H_2O$ was prepared by adding 6 g of zinc acetate dihydrate into 50 ml distilled water at room temperature in a glass beaker under magnetic stirring.

B.   0.25 mm of different studied surfactants were mixed with the previous prepared aqueous solution.

C.   In order to reduce the previous zinc acetate salt, 50 mm of sodium hydroxide (NaOH) was added drop wise until reaching to the required pH nearly 9 of the final solution.

D.   The obtained mixture was heated to various temperatures and continued with these temperatures until the reaction completed.

E.   The obtained white powders were washed several times with distilled water and absolute ethanol to remove any residual salts, and centrifuged at 6000 rpm for 30 minutes to precipitate the Nano-zinc oxide.

F.   Finally, the resultant powders were dried at 60°C under air atmosphere overnight.

2. Characterization of ZnO Nano powders

The physical properties of the synthesized Nano-zinc oxide with different morphological structures were investigated using different techniques. The morphological structure and the chemical compositions of the ZnO Nano powder were examined.

A). X-ray Diffraction Analysis (XRD)

X-ray powder diffractometry was carried out using (Schimadzu 7000).

B). Scanning Electron Microscopy (SEM)

The scanning electron microscope is based on scanning a finely focused electron beam across the surface of a specimen. The surface of the prepared ZnO Nano powders was scanned with scanning electron microscope to investigate the homogeneity of the Nano powder and to measure the dimensional structures of the different architecturing prepared powders.

3. Enrichment of Listeria Reference (ISO 11290-2/A1-2004)

According to ISO 11290-2/A1-2004 Listeria reference strain enriched on 10 ml Half-Frazer base then incubated for 1 hr ±5 min at 20 ± 2 ºC, then0.1 ml spread onto ALOA agar plates & Incubated for 24 hrs ±3hrs at 37 ºC ±1 ºC (If necessary incubate an additional 24 hrs ±3 hrs, Enumerate characteristic colonies (turquoise-blue with opaque halo), then isolate cells to treat every gram of the samples with 120 cells.

4. Preparation of samples(ISO 17604:2003)(ISO 11290-2/A1-2004)

A representative part of 25 gm. taken from Each sample of 250 gm. in weight under sterilized conditions according to (ISO 17604:2003) to prepare 240 samples of RTE meat

products (Frankfurter, Salami, Basterma& Luncheon)(60 sample of each product),
divided in 12 groups, Three groups for each product ,Each group consists of 20 samples
(1st control group, 2nd group & 3rd group), the control group samples of each product was
not treated with Zn O NPs , the 2nd group samples of each product was treated with 60 ppm
(60μg/gm) of Zinc oxide Nanoparticles & the 3rd group samples of each product was
treated with 90 ppm (90μg/gm) of Zinc oxide Nanoparticles , then listeria monocytogenes
prepared to be inoculated in the All groups of samples with the same concentration 120
cells/g (120 CFU/g), then each sample (25g) added to 225 ml Half-Fraser base without
antibiotics for 1 hr ±5 min at 20 ± 2 ºC, then 0.1 ml spread onto ALOA agar plates and
Incubated for 24 hrs ±3hrs at 37 ºC ±1 ºC (If necessary incubate an additional 24 hrs ±3 hrs,
Enumerate characteristic colonies (turquoise-blue with opaque halo) to evaluate the
antibacterial activity of Zn O NP on listeria monocytogenes.

Results & discussion:

According to the obtained Results, There was no any inhibition in the control groups which
was not treated by any concentration of Zn O NPs, on opposite there was a significant
increase in the number of CFU/g of Listeria monocytogenes in samples of control groups of
all RTE Meat products, this increase due to the growth and the multiplication of the micro-
organisms.

In the 2nd group of each product which was treated with Zn O NPs in concentration of
60 ppm, there was some inhibition of the no. of CFU/g, while in the 3rd group which was
treated with Zn O NPs in concentration of 90 ppm, there was a significant inhibition of no of
CFU/g of L. monocytogenes more than which was in the 2nd group. The inhibition of Listeria
monocytogenes increases when the concentration of Zn ONPs increases . Table (2).

During the present study, two different concentrations of Zinc oxide nanoparticles were
tested to find out the best concentration that can have the most effective antibacterial
property against the Listeria monocytogenes. Our data is in accordance with (Saleh et al.,
2014), bacterial inhibition depends upon the concentrations of Zinc oxide nanoparticles. It
reflects that Zinc oxide nanoparticles have an excellent antibacterial effect and potential in
reducing bacterial growth for practical applications specially on Listeria monocytogenes.

7

Table (2):Efficacy of different concentrations of Zn O NPs in reduction of L. monocytogenes colonies inoculated in RTE meat products.

(Mean ± SD)CFU/g

RTE Meat Product	1st control group	2nd group	3rd group
	Zn O NPs Added None	Zn O NPs Added 60 ppm (60µg/gm.)	Zn O NPs Added 90 ppm (90µg/gm.)
	Initial concentration of L. monocytogenes(1.2×10^2)	Initial concentration of L. monocytogenes (1.2×10^2)	Initial concentration of L. monocytogenes(1.2×10^2)
Frankfurter	$4.63 \times 10^2 \pm 0.11 \times 10^2$CFU/g [a]	$6.8 \times 10^1 \pm 0.52 \times 10^1$CFU/g [b]	$5 \times 10^1 \pm 0.68 \times 10^1$CFU/g [c]
Salami	$4.7 \times 10^2 \pm 0.09 \times 10^2$CFU/g [a]	$7.1 \times 10^1 \pm 0.47 \times 10^1$CFU/g [b]	$5.2 \times 10^1 \pm 0.57 \times 10^1$CFU/g [c]
Basterma	$5.09 \times 10^2 \pm 0.09 \times 10^2$CFU/g [a]	$7.7 \times 10^1 \pm 0.63 \times 10^1$CFU/g [b]	$5.6 \times 10^1 \pm 0.57 \times 10^1$CFU/g [c]
Luncheon	$4.99 \times 10^2 \pm 0.13 \times 10^2$CFU/g [a]	$7.5 \times 10^1 \pm 0.58 \times 10^1$CFU/g [b]	$5.3 \times 10^1 \pm 0.59 \times 10^1$CFU/g [c]

- Means with similar litter are not significantly different ($p \leq 0.05$)

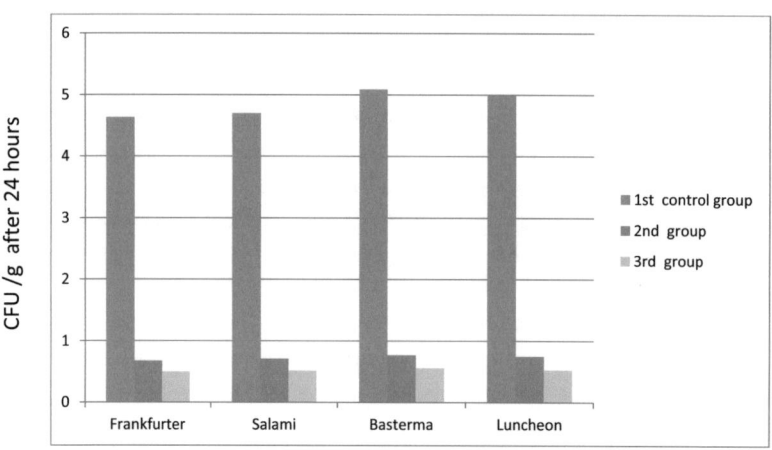

Figure (1): CFU/g In All Groups of Samples of RTE meat products.

Conclusion:

Actually in this study the effect of Zn O NPs on L. monocytogenes in RTE meat products was positive and satisfactory because the inhibition of Listeria monocytogenes increased when the concentration of Zn ONPs increased.

there was some inhibition of the no of CFU/g in the products which was treated with Zn O NPs in concentration of 60 ppm, While there was a significant inhibition in no of

L. monocytogenes with using of maximum concentration of Zn O NPs 90 ppm .It reflects that, Zinc oxide nanoparticles have an excellent antibacterial effect and potential in reducing bacterial growth for practical applications specially on Listeria monocytogenes.

to achieve a complete inhibition of CFU/g depending on Zn O NPs as a meat preservative we need many researches to act on this point, to studying the effect of increasing Zn O NPs concentration, and to increase the penetration power of Zn O NPs in RTE meat products .

References:

Allerberger, F. & Wager, M. (2010). Listeriosis: a resurgent food borne infection. Clinical Microbiology and Infectious Diseases, 16, 16-23.

Anonymous. (2007). Preliminary Food Net data on the incidence of infection with pathogens transmitted commonly through food-10 states, 2006. Morbidity and Mortality Weekly Report, 56, 336-339.

Bourry, A. &Poutrel, B. (1996). Bovine mastitis caused by L. monocytogenes: Kinetics of antibody responses in serum and milk after experimental infection. Journal of Dairy Science, 79, 2189-2195.

Buzby, J. (2002). Older adults at risk of complications from microbial food borne illness. Food Review, 25, Issue2, 30-35.

Clement JL, Jarrett PS.Met Based Drugs.1994, 1, 5-6, 467-82

ISO/IEC 11290-1 : Microbiology of food and animal feeding stuffs—Horizontal method for the detection of Listeria monocytogenes.

ISO/IEC 11290-2 : Microbiology of food and animal feeding stuffs—Horizontal method for enumeration of Listeria monocytogenes.

ISO/IEC 17604, (2003) :Microbiology of food and animal feeding stuffs- Carcass sampling for microbiological Analysis. Page 6-7.

ISO (7218:2007): Microbiology of food and animal feeding stuffs, General requirements and guidance for microbiological examinations.

Jin, T.; Sun, D.; Su, J.Y.; Zhang, H.andSue, H,J. (2009): "Antimicrobial efficacy of zinc oxide quantum dots against Listeria monocytogenes, Salmonella Enteritidis, and Escherichia coli O157:H7", J. Food Sci., 74(1), 46-52.

Jones, N., B. Ray, K. T. Ranjit, and A. C. Manna. 2008. Antibacterial activity of ZnO nanoparticle suspensions on a broad spectrum of microorganisms. FEMS Microbiol.Lett. 279:71–76.

King, J., Black, R., Doyle, M., Fritsche, K., Halbrook, B.,Levander,O., Meydani, S., Walker, W., Woteki,C. (2000).Food borne illnesses and nutritional status: A statement from an American Society for nutritional sciences working group. The Journal of Nutrition,130, 2613-2617

Kneifel, W. & Pacher, B. (1993) : An X-glu based agar medium for the selective enumeration of Lactobacillus acidophilus in yoghurt-related milk products. International Dairy Journal, 3, 277 291.

Kühn, H; Wonde, B; Rabsch, W.and Reissbrodt, R. (1994): Evaluation of Rambach agar for detection of Salmonella Subspecies I to VI.Applied and Environmental Microbiology, 60, 749 751.

Liu, Y., et al. 2009. Antibacterial activities of zinc oxide nanoparticles against Escherichia coli O157:H7. J. Appl. Microbiol. 107:1193–1201.

Manafi, M. (1996) : Fluorogenic and chromogenic enzyme substrates in culture media and identification tests. International Journal of Food Microbiology, 31, 45 58.

Manafi, M; Kneifel, W and Bascomb, S. (1991): Fluorogenic and chromogenic substrates used in bacterial diagnostics. Microbiological Reviews, 55, 335 348.

McGlauchlen,K. &Vogel,L.(2003).Ineffective humoral immunity in the elderly. Microbes and Infection, 5, 1279-1284.

McLauchlin, J. 1996. The role of the public health laboratory service in England and Wales in the investigation of human listeriosis during the 1980's and 1990's.Food Control 7:235-239.

Mead, P., Slutsker,L, Dietz, V., McCaig,L.,Bresee, J., Shapiro, C., Griffin, P., Tauxe, R. (1999).Food-related illness and death in the United States. Emerging Infectious Diseases, 5(5), 607-625.

Nikolich-Zugich, J. (2008). Aging and life-long maintenance of T-cell subsets in the face of latent persistent infections. Nature Reviews Immunology, 8, 512-522.

Ottaviani, F; Ottaviani, M and Agosti, M. (1997a): Esperienzasuun agar selettivo e differentiale per Listeria monocytogenes. Industrie Alimentari, 36, 1 3.

Ottaviani, F; Ottaviani, M and Agosti, M.(1997b): Differential agar medium for Listeria monocytogenes. In: Quimper Froid Symposium Proceedings, p. 6. A.D.R.I.A. Quimper, France, 16–18June 1997.

Ryser, E., &Marth E. (Eds.) (2007). Listeria, Listeriosis, and Food Safety, 3rd Ed.New York: Marcel Dekker.

Saleh,E.A .,Alkamary,M.A .,Hafez,E.E and Elfiky S.E. 2014 : effect of zinc oxide nanoparticles on Listeria Monocytogenes in meat products, 1st international conference on Environmental hazards on food safety, Zagazig University .

Scharff, R. (2010). Health-related costs from food borne illness in the United States. Retrieved August 1,2010from http://www.producesafetyproject.org/reports?id=0008

Slotwiner-Nie, P. (2001). Infectious diarrhea in the elderly. Gastroenterology Clinics of North America, 30, 625-635

SmithJ.(1998). Food borne illness in the elderly.Journal of Food Protection, 61, 1229-1239.

Swaminathan, B.&Gerner-Smidt, P. (2007). The epidemiology of human listeriosis. Microbes and Infection, 9, 1236-1243.

Wong, S., Street, D., Delgado, S.,&Klontz, K. (2000). Recalls of foods and cosmetics due to microbial contamination reported to the U.S Food and Drug Administration. Journal of Food Protection, 63, 1113-1116.